How well do you know your dinosaurs?

A Debut

MIR GILL

Book of Dreams, Inc
Chicago, IL

This book is non-fiction. Facts are a product of the author's research. Any resemblance to another work is entirely coincidental.

Copyright © 2026 by Mir Gill

Publisher Book of Dreams, Inc

Visit us at **www.bookofdreams.us**

Library of Congress Cataloging-in-Publication Data Names: Gill, Mir, 2026

Classification: LCCN 2025925263 | ISBN 978-1-7349423-5-4

Printed in the United States of America

Table of Contents

The Dinosaur Timeline 3

The Triassic Period 5

The Jurassic Period 7

Allosaurus 9

Stegosaurus 11

Brachiosaurus 13

The Cretaceous Period 15

Tyrannosaurus Rex 17

Triceratops 19

Ankylosaurus 21

Pachycephalosaurus 23

Extinction 25

Conclusion 27

**If we traveled back in time, we wouldn't see this scene, which is merely an illustration. These dinosaurs didn't live at the same time or in the same place.*

The Dinosaur Timeline

Dinosaurs lived from 252 to sixty-six million years ago, a span of 186 million years across three time periods: the Triassic, Jurassic, and Cretaceous.

1. **The Triassic period** lasted fifty million years from 252 to 201 million years ago. The period ended with a mass extinction of the Triassic dinosaurs.
2. **The Jurassic period** lasted fifty-six million years from 201.4 million to 145 million years ago.
3. **The Cretaceous period** lasted nearly eighty million years, forming the longest geological period in the phanerozoic eon.

Note that dinosaurs didn't all exist at the same time. For example, *Stegosaurus* never saw *T. rex* because they lived in non-overlapping periods and in different regions of the planet.

Let's looks at dinosaurs from each period in depth next.

The Triassic Period

The first dinosaurs roamed the Earth during the Triassic period, 252 million years ago. The largest dinosaurs at this time were about the size of a horse. These included *Eoraptor, Herrerasaurus, and Coelophysis.*

When the supercontinent of **Pangea** split up, lava from the volcanoes filled in the cracks, which resulted in a mass extinction.

Dinosaurs grew to astonishing sizes during the Jurassic period.

The Jurassic Period

Two hundred one million years ago, the Jurassic period began, and dinosaurs grew to humongous sizes—some reaching nearly four dozen feet. At this time, they firmed their place at the top of the food chain.

Dinosaurs of this period included the tall *Brachiosaurus*, the spiky *Stegosaurus*, as well as *Allosaurus*, *Diplodocus*, and *Apatosaurus*.

Allosaurus *means "different lizard."*

Allosaurus

Allosaurus lived in the late Jurassic period. Its huge skull included crests over its eyes, a feature which uniquely distinguished it.

An adult measured twenty-eight to thirty-five feet and weighed one to two tons.

Allosaurus varied from *T. rex*. For example, *T. rex* had big, D-shaped teeth to crush bones while *Allosaurus's* teeth were small and flat used for tearing flesh.

It may have used its head to slam into prey.

In groups, not packs like wolves, *Allosaurus* lived in what is now Utah and Colorado in North America. So what's the difference between packs and groups? Groups are smaller in size than packs.

Allosaurus preyed on *Stegosaurus* and young **sauropods**. Like all **carnivore** dinosaurs, *Allosaurus* was a **bipedal** predator.

Stegosaurus *means "roof lizard" because its back was roofed with plates.*

Stegosaurus

In the late Jurassic period, 155 million years ago, lived the *Stegosaurus*.

Hunted by *Allosaurus, Stegosaurus* defended itself with its tail spikes while the plates on its back helped cool it down.

Did you know its brain was the size of a walnut?

Around twenty-six to thirty feet (nine meters) long and weighing up to seven tons, *Stegosaurus* had shorter front legs and longer back legs.

It was among the earliest dinosaurs to possess cheeks used for chewing and storing food.

Being an **herbivore**, it ate plants—but also one unusual item: rocks.

Stegosaurus chewed on rocks not for nutrition, but to help with its digestion.

Brachiosaurus *walked making thunderous earthquake sounds.*

Brachiosaurus

Brachiosaurus was a **sauropod** (a type of dinosaur with long legs and tall necks, about thirty feet high).

It lived in the late Jurassic period, in the same region as *Stegosaurus*.

Brachiosaurus lived peacefully.

Weighing thirty-five to sixty tons (as heavy as ten elephants), *Brachiosaurus* was eighty feet long, making it a tall **herbivore** that could reach the highest trees no other peer could.

The height had other benefits as well.

Because of their huge size, they rarely got attacked by the predators.

It would lift its two front legs and land its predators. Then, it would blare thunderously, scaring them away.

Its name means "arm lizard" because its front legs are longer than its back legs.

The Cretaceous period was the last before the asteroid hit the planet.

The Cretaceous Period

The Cretaceous period was sixty-six million years ago, boasting the biggest **carnivores**: *Carcharodontosaurus, T. rex, Edmontosaurus, Pachecephosaurus, Triceratops,* and *Giganotosaurus.*

Tyrannosaurus Rex (T. rex) *means "tyrant lizard king."*

Tyrannosaurus Rex

Tyrannosaurus rex, you know as *T. rex*, lived at the very end of the Cretaceous period, sixty-eight to sixty-six million years ago.

T. rex was forty-two feet long, reaching up to twenty feet. Compare that to the TALLEST human's height of eight feet!

It hunted *Triceratops* and *Edmontosaurus*.

T. rex lived on an island continent called Laramidia, what is now North America. *T. rex* had a much wider range of habitat compared to the other tyrannosaurids as its fossils are found in a variety of geological regions.

Fun fact, *T. rex* had the biggest jaw of any dinosaur. Its jaw's weight was balanced by a heavy tail.

The biggest and most complete *T. rex* (called "Sue") was found in South Dakota and named after Sue Hendrickson, who discovered it. It's now proudly displayed in the Chicago's Field Museum.

*Tri means **three**; cera means **horns**; top stands for **face**. So* Triceratops *means **three horned face**.*

Triceratops

One of the most popular dinosaurs ever to live was *Triceratops*. It could *beat the mighty* T. rex.

Sixty-seven million years ago, *Triceratops* lived with *T. rex* in the late Cretaceous period.

They belonged to the ceratopsian family with distant cousins like *Styrocosaurus*, *Albertasaurus*, *Kosmosaurus*, and *Eleinosaurus*.

They could be found in Hell Creek, North America, and in the western America.

About thirty feet long, weighing up to twelve tons, *Triceratops* had one of the biggest skulls—as long as eight feet.

It had 800 teeth arranged in stacked columns and five layers, plus a groove that helped it chew tough plants.

Fun fact, most mistake the frill was used for safety, but *Triceratops* employed it for mating.

Ankylosaurus *means "fused lizard" because many of its bones were fused.*

Ankylosaurus

During the late Cretaceous period, sixty-seven to sixty-eight million years ago, *Ankylosaurus* lived in North America (specifically in Wyoming and Montana).

Thirty feet long and weighing about four to seven tons, *Ankylosaurus* used the spikes on its back along with its clubbed tail, domed or balled, to defend from predators.

Its predators—*T. rex, Dakotaraptor, Albertosaurus,* and *Dromaeosaurus*—were impeded not just by the spikes but also by its heavy crust, which they couldn't bite through, so they had to improvise. They attacked *Ankylosaurus* by flipping it over.

Its tail alone weighed approximately 990 pounds. That's almost as heavy as the world's heaviest person!

Ankylosaurus was an **herbivore**.

Pachycephalosaurus *was known for its agility.*

Pachycephalosaurus

(PACH EE SEPH ALL A SAUR US)

Pachycephalosaurus lived in the late Cretaceous period seventy to sixty-six million years ago.

It's known for its domed head, which it used for defending, playing (smashing heads together), and mating. With age, the head would get bigger and less flat. Composed of bone and blood vessels, it was about ten inches deep while *Pachycephalosaurus* itself was about fifteen feet long and weighed a ton.

Some scientists believe it had feathers. For defense, it depended on its agility.

Did you know why *Pachycephalosaurus* was so fast? Despite being an **herbivore**, it walked on two legs, which made it **nimble**.

It ate leaves, fruits, soft plants, and possibly insects. No wonder because it didn't have sharp teeth or a big jaw. *Pachycephalosaurus* was small but aggressive. If it felt threatened, you better run.

Extinction

The dinosaurs' rein expired suddenly at the end of the Cretaceous period, concluding the Mesozoic era.

How did they go extinct?

Was it from an asteroid or a volcano? Guess what? It was both! An asteroid started a volcano, which also spawned **tsunamis** and **hurricanes**. A deep crater at the Gulf of Mexico, half underwater and half on land, serves as proof today.

Conclusion

As you can see, dinosaurs are more interesting and complex than just some large animals who lived millions of years ago. They once enjoyed being at the top of the food chain before getting annihilated.

These fascinating creatures ruled the Earth during the AGE OF REPTILES!

ABC's of dinosaurs

A – Apatosaurus

B – Brachiosaurus

C – Corythosaurus

D – Deinonychus

E – Einiosaurus

F – Fabrosaurus

G – Gallimimus

H – Hadrosaurus

I – Iguanodon

J – Jaxartosaurus

K – Kentrosaurus

L – Lambeosaurus

M – Megalosaurus

N – Nodosaurus

O – Ornithomimus

P – Parasaurolophus

Q – Qantassaurus

R – Rhabdodon

S – Stegosaurus

T – Tyrannosaurus

U – Utahraptor

V – Velociraptor

W – Wannanosaurus

X – Xenotarsosaurus

Y – Yangchuanosaurus

Z – Zigongosaurus

Glossary

- **Herbivore** is a plant eater.

- **Carnivore** is a meat eater.

- **Nimble** means fast.

- **Sauropod** are dinosaurs with huge necks and legs.

- **Bipedal** means walking on two feet like humans.

- **Tsunami** is an ocean wave caused by a large earthquake, volcanic eruptions, landslides etc.

- **Hurricane** is a tropical storm with winds greater than seventy-four miles originating from the ocean.

- **Pangea** is the ancient supercontinent that joined nearly all of Earth's landmasses into one giant continent, surrounded by a single ocean called Panthalassa.

About the Author

Amazed by dinosaurs, **Mir Gill** wrote this book when he was just a nine-year-old boy, dreaming about becoming a paleontologist.

He reads daily, especially books filled with facts about dinosaurs, fossils, and paleontology like *The Age of Dinosaurs* and *The Rise of Dinosaurs* by Steve Brussate.

With an older brother and sister, he lives in Chicago. He comes from a line of authors, among them his mother, Mars D. Gill, and his great-grandfather, Surjan Singh Gill.

www.ingramcontent.com/pod-product-compliance
Lightning Source LLC
LaVergne TN
LVRC091352060526
838200LV00016B/370